Maria Vitória Pereira Vicente

Ethnobotanical Recovery of Medicinal Plants

Maria Vitória Pereira Vicente

Ethnobotanical Recovery of Medicinal Plants

Survey of students at IFSULDEMINAS - Campus Muzambinho - MG

ScienciaScripts

Cover image: www.ingimage.com

This book is a translation from the original published under ISBN 978-613-9-64790-3.

Publisher:
Sciencia Scripts
is a trademark of
Dodo Books Indian Ocean Ltd. and OmniScriptum S.R.L publishing group

120 High Road, East Finchley, London, N2 9ED, United Kingdom
Str. Armeneasca 28/1, office 1, Chisinau MD-2012, Republic of Moldova, Europe
Printed at: see last page
ISBN: 978-620-7-78653-4

CONTENTS

SUMMARY

The World Health Organisation points out that 80% of humanity does not have access to primary health care, which is why many people use medicinal plants to cure various ailments. However, traditional folk knowledge is passed on in unrecorded ways. Medicinal plants are an important tool for promoting health, both for rural and urban populations. The aims of this research were to retrieve and record traditional folk knowledge about medicinal plants among students at the IFSULDEMINAS Campus Muzambinho - MG. Four hundred students were interviewed using a semi-structured questionnaire and each answered fifteen questions related to the use of medicinal plants. The most commonly cited species were mint *(Menta x villosa* Huds) (32%), chamomile (*Matricaria chamomilla) (*23%) and lemon balm *(Cymbopogon citratus (*DC). Stapf (20%). The interviewees reported that they had received this knowledge from their parents and grandparents. The most common illnesses treated with medicinal plants reported by IFSULDEMINAS Campus Muzambinho - MG students are colds and flu, migraines, anxiety and stress, and stomach ache. The work contributes to the traditional popular revival of medicinal plants and their registration, thus preventing the extinction of this knowledge and helping to ensure that this tradition does not end, transmitting correct knowledge of how to use medicinal plants and the ideal preparation of each part of the herb.

Key words: Ethnobotany; Popular knowledge; Medicinal plants.

CHAPTER 1

INTRODUCTION

In ancient times, people used plants for their livelihood, and gradually discovered which species were medicinal and which were toxic, thus giving rise to the popular culture of using medicinal plants.

Data from the World Health Organisation show that around 80% of the world's population uses some kind of herb to relieve painful or unpleasant symptoms (MARTINS et al., 2003).

Much of what has been studied and is currently known by everyone about treatments with plants originated in the culture of popular knowledge. Scientific knowledge is an evolution of all the methods used in the past with medicinal plants, and this method is still very present in our society, often due to the high price of synthetic medicines and the ease of obtaining them.

Through popular knowledge, various illnesses have been cured through the use of medicinal herbs that were selected for the ailment.

This led several communities to pass on this knowledge orally, which was later catalogued and archived, thus allowing several generations to have access to other forms of treatment that could be an alternative to expensive treatment with synthetic materials. There have been changes over time to this tradition, which has often been lost, but there are still several people who try to keep the knowledge alive. This has led to an interest in recovering this knowledge, along with its management, in order to record informal learning, thus valorising traditional folk medicine.

Every society has information and traditions about the environment that surrounds it, which makes it possible for the environment to interact with society so that it can take advantage of the natural materials available for its human needs. The relative knowledge of the universe of medicinal plants and their use brings society into contact (AMOROZO, 1996).

Accumulating empirical knowledge about how plants work has been passed down from ancient times to the present day, where the use of medicinal plants has become a popular medicinal practice (DORIGONI et al., 2001).

The use of medicinal plants is strongly present in the popular culture of all societies and is passed down from parents to children throughout human existence. This knowledge is found among traditional (DIEGUES, 1996) and/or contemporary peoples, and from what has been observed, it tends to diminish or even disappear when it is subjected to the action of modernity (GUARIN NETO et al., 2000).

In general, communities pass on their knowledge through unrecorded information exchanged by individuals, and this can lead to information being passed on in a mistaken or altered way. Therefore, rescuing this knowledge and its therapeutic techniques is a way of recording an informal way of learning that contributes to the valorisation of traditional folk medicine, as well as generating information about the health of the local community (PILLA et al., 2006).

Ethnobotany not only rescues plant species used as remedies, but also values the popular knowledge of communities (AMOROZO et al., 1996) and, through studies in this area, medicinal plants can have therapeutic actions.

For a long time, representations of traditional medicinal treatments were considered exotic and ineffective, characteristic of backward societies and destined to disappear with time and the implementation of Western medicine (BUCHILLET, 1991), but these treatments are still effective and cure many diseases today.

Despite the widespread dissemination of knowledge about medicinal plants, their preparation is not always carried out correctly and this leads to the loss of the active ingredient, followed by a drop in the plant's expected effect on the body. Many users carry out the procedures in a contrary and ineffective way, and some forms of improper preparation cause damage to the final result. Many people still believe that they are devoid of toxic effects and that they do not cause any damage to health, which is not true (AGRA et al., 2007); (PEREIRA et al., 2005), a fact that was worrying throughout the research, as the vast majority of interviewees do not know how to differentiate toxic plants from healthy plants.

And even though they are natural, the consumption of these substances must be regulated, as exaggeration can lead to future health problems and the mixture of various herbs can cause unpredictable reactions.

Traditional folk medicine has been making an increasing contribution to the human sciences, due to the amount of empirical medical knowledge and practices, which are influenced by the culture and economy in which they are inserted in a society (CAMARGO, 1976).

Plants have chemical components that are responsible for their therapeutic action and which can change according to the location and the characteristics of cultivation and processing to which they are subjected, making them unfit for consumption (FRANCO et al., 2007); (SILVA et al., 2003). In order to take advantage of all the benefits of medicinal plants, it is necessary to prepare and choose the parts to be used in such a way that the plant's active ingredients remain unchanged after preparation and are able to exert the desired therapeutic action.

The aim of this study was to carry out an ethnobotanical survey of the medicinal plants used by students at the IFSULDEMINAS Campus Muzambinho - MG, in order to retrieve and record traditional folk knowledge about them.

CHAPTER 2

LITERATURE REVIEW

Medicinal plants symbolise the medical raw material used by traditional or non-Western medicine, with folk medicine using the largest number of different species (HAMILTON, 2003).

Popular knowledge is developed by various groups from different cultures who still live with nature, observing it and exploring its potential, thus keeping this heritage alive and growing through systematic and constant experimentation (ELISABETSKY, 1997).

Medicinal plants have always been used, and in the past they were the main known means of treating the population.

This knowledge is maintained through oral tradition, and because of this, little information has been proven about its beneficial and harmful effects (OLIVEIRA; ARAÚJO, 2007).

The plants used are based on popular knowledge and lack scientific knowledge. However, the pharmacological properties used and advertised lack scientific validation, as they have not been proven in clinical trials. Furthermore, there is little knowledge about the constituents with pharmacological activity and about possible interactions with the numerous molecules present in the plant (TUROLLA; NASCIMENTO, 2006).

CHAPTER 3

MATERIAL AND METHODS

The aim of this research was to carry out an ethnobotanical study in which data was collected on the traditional use of medicinal plants by students at the IFSULDEMINAS Campus Muzambinho - MG.

The data obtained was analysed using the free software LibreOffice 6.0.5 and Google Drive, which was used to produce the statistics and graphs, along with the questionnaire.

The research was carried out at the IFSULDEMINAS Campus Muzambinho - MG, located in the south of Minas Gerais, between March/2017 and June/2017. The population of Muzambinho is 20,430 inhabitants, with an estimated 21,012 inhabitants in 2014, and it has an area of 409.948 km^2 whose biome is Atlantic Forest (IBGE, 2014).

The students who took part in the research were volunteers and chosen at random. The interviews took the form of a semi-structured questionnaire with open and closed questions on the premises of the IFSULDEMINAS Campus Muzambinho - MG. The questionnaire consisted of questions about the medicinal species most frequently used by the interviewee and their family, who taught them how to use them and how they were prepared. The results obtained were subjected to descriptive statistical analysis using relative frequency.

The researcher used dialogues to direct the conversation in order to clarify doubts and tried to develop a structured questionnaire with open and closed questions (ALBUQUERQUE & LUCENA, 2004). The time spent on each interview varied from ten minutes to half an hour, depending on the interviewees' knowledge of medicinal plants.

The interviews were carried out full-time with students from all the courses, from integrated technical to higher education. It was observed that many of the students on the Integrated Agricultural Technician course do not know how to differentiate toxic plants from non-toxic plants, even though their curriculum includes subjects on botany.

However, not only do they not know how to identify these types of plants, which raises concerns about their use of certain plants.

During the interviews, people recounted their knowledge of plants, who they had learnt to use them from, and the illnesses that their parents and grandparents treated with medicinal plants, because at the time not everyone had access to industrialised medicines and because of the belief they passed on, they believed that plants were beneficial to health.

The aim was to emphasise all the ethnobotanical data provided, such as the most commonly used plants, in which situations to use which plant and which part and how to prepare it, consumption, purpose, which age group they think uses the herbs the most. In addition to identifying how they obtained their knowledge of the plants, their level of education and their age.

The plants were identified using the common names commonly known by the students, and then compared with the scientific names, using appropriate material such as books and internet searches on medicinal plants and when they should be used.

The questionnaire developed for the survey and to gather information is attached as Annex I, which contains all the information about the reason for having chosen to develop it.

Google Maps was also used to obtain satellite images to illustrate the city in which IFSULDEMINAS - Campus Muzambinho and the Institute itself are located.

Below, satellite images of the city of Muzambinho - MG and the Federal Institute of Education, Science and Technology of Southern Minas Gerais - Campus Muzambinho.

Satellite image of the Federal Institute of Education, Science and Technology of Southern Minas Gerais - Campus Muzambinho

Satellite image of the city of Muzambinho - Minas Gerais.

CHAPTER 4

RESULTS AND DISCUSSIONS

This study interviewed four hundred students, 76% of whom said they used medicinal plants frequently, which shows that most of them use medicinal species and 93% received this knowledge from their parents and grandparents, which is in line with the study by ALMEIDA et al. (2009), where the use of medicinal plants in the region is passed down from generation to generation and around 74% of the informants in the municipality of the study use them.

The students taking part in the research were aged between fifteen and twenty-five and they mentioned three different species of medicinal plants that they use the most, the ones mentioned were *Mint (Menta x villosa* Huds) (32%), Chamomile (*Matricaria chamomilla) (23%)* and Lemon Balm *(Cymbopogon citratus (*DC) (20%). Stapf.

Although the interviewees were young, they showed a great deal of knowledge about the healing power of plants, which is different from the work of RODRIGUES et al. (2002) in Luminárias, where the population of the municipality that had the greatest knowledge about the power of plants was on average 60 years old. Some reported preferring to use medicinal plants rather than industrialised medicines.

It was found that 53 per cent of those interviewed use the species because they are ill and 47 per cent say they use medicinal herbs because they have already acquired the habit, since those who say they have already acquired the habit use them more than once a week, while those who use them only when they are ill use herbs they are already used to.

Another aspect of the survey was the credibility that students had in the healing of medicinal species, with 77 per cent believing that they cure various pathologies and only 23 per cent reporting that medicinal plants only heal in combination with other medicines, as one interviewee reported, that family members used medicinal plants to help treat cancer in the family, and that because they often didn't see the medicines having an effect and because of the tradition that their grandmother brought with her, which she had received from her parents, they had the idea of trying to use teas and home-made remedies for an effective and faster recovery. The combination of herbs and other medicines was successful and the sick member of the family was cured of the illness. There was another report that a more distant relative of one of the interviewees used medicinal plants alone to treat migraines and sinusitis.

With regard to the toxicity of medicinal species, a worrying observation was made, as 77 per cent of those interviewed said they did not know how to identify which plants were toxic, using only medicinal herbs that friends and relatives recommended. This is a very serious situation, as the risk of poisoning can often occur due to a lack of knowledge about the correct dosages for using medicinal species and not recognising toxic species. This is a worrying fact, which generated a discussion, because as IFSULDEMINAS - Campus Muzambinho originated as an agricultural college, many of the students on nature-related courses don't know how to identify toxic plants and most of them have subjects involving botany on their curriculum.

Many plants produce chemical substances that act on the body in a toxic way and not knowing how to identify toxic plants can lead to serious problems.

Students from different courses also reported that to find out whether a plant is toxic or not, they break the stem or leaf of the plant, and if there is a white liquid present, the plant is toxic. This method is widely used by several of the interviewees and their acquaintances.

The most commonly used methods for preparing the species were tea by decoction (54.5 per cent) and tea by infusion (42.5 per cent), but the interviewees used other methods. Ointments were mentioned, among them the most commonly cited was arnica, which has a therapeutic effect on muscle distension, and poultices were also mentioned to help treat various health problems.

When asked which age group the interviewees thought used medicinal herbs the most, 56 per cent said children, because children usually can't take several medicines and so doctors and family members recommend teas. This was followed by young people and then the elderly, while adults came last, since most adults often don't even go to hospital or seek help when they are ill due to lack of time.

This led us to discover that 67 per cent of those interviewed, when they have a cold or illness, prefer to stay at home and wait for it to pass, and if there are complications or it gets worse, it's time to seek medical help. And 33 per cent already seek medical help when they feel unwell, so that there are no complications.

The knowledge of medicinal plants and their proper use is passed down in a traditional way, like an heirloom, from father to son, and few people seek to learn about the subject via the internet or any media or even by consulting a friend. Family members are responsible for keeping this culture alive, as the knowledge of the use of medicinal herbs comes from ancestors.

Regarding health problems, it was found that the flu is the most common for most of the interviewees, followed by migraines. And 68 per cent of the cases are solved by the parents rather than the interviewees themselves.

The graphs below show the results of all the questions in the questionnaire applied to the interviewees:

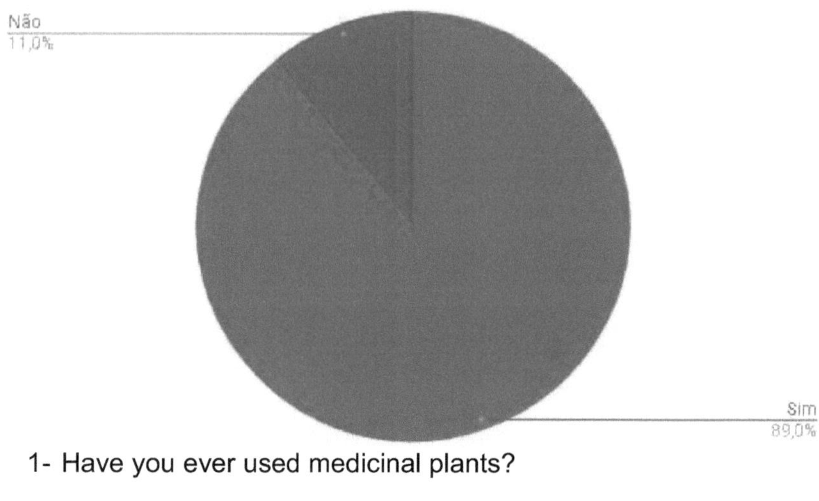

1- Have you ever used medicinal plants?

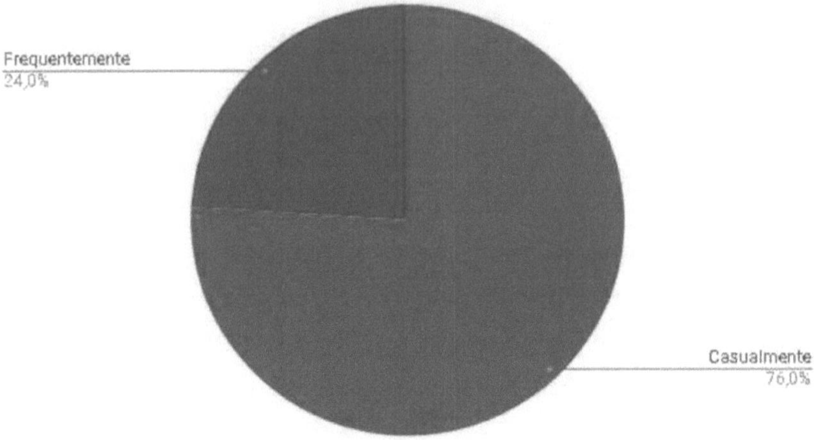

2-How often:

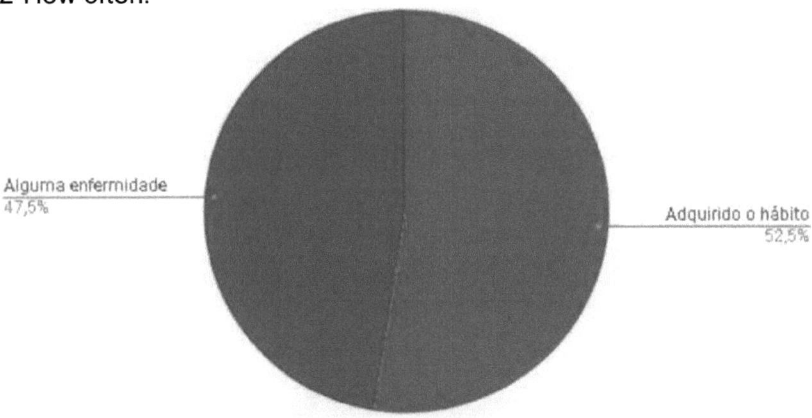

3-What do you use medicinal plants for?

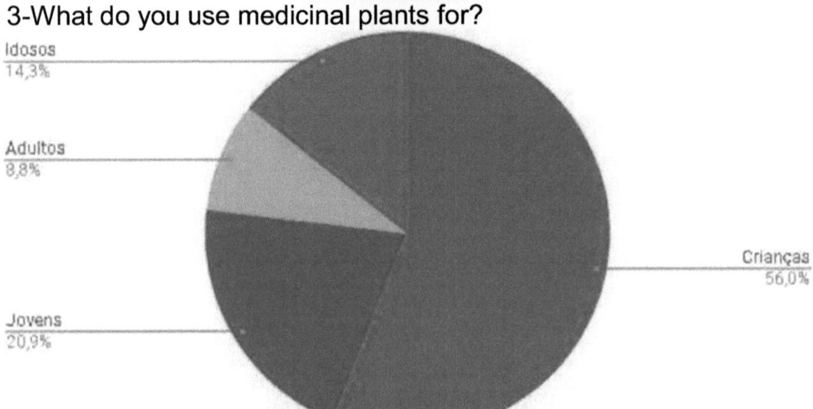

4-For you, which age group predominates in the use of Medicinal Plants:

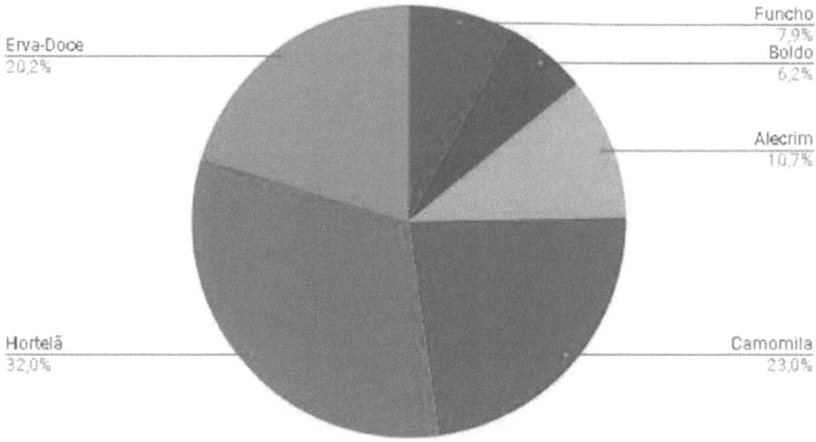

5-What species do you use?

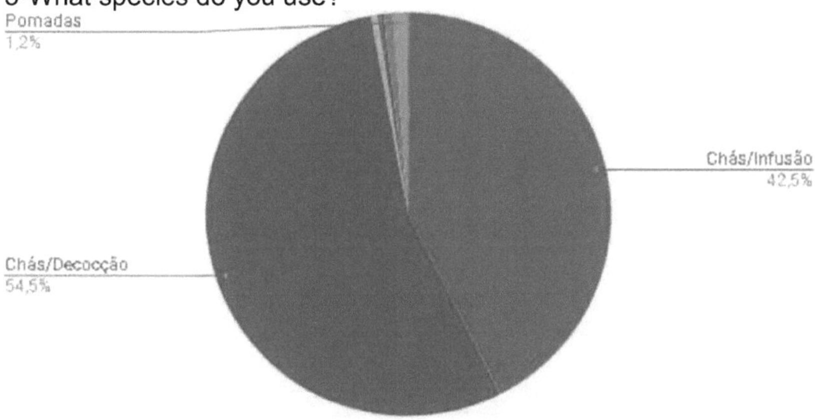

6-You have used or are using these plants medicinally in what way?

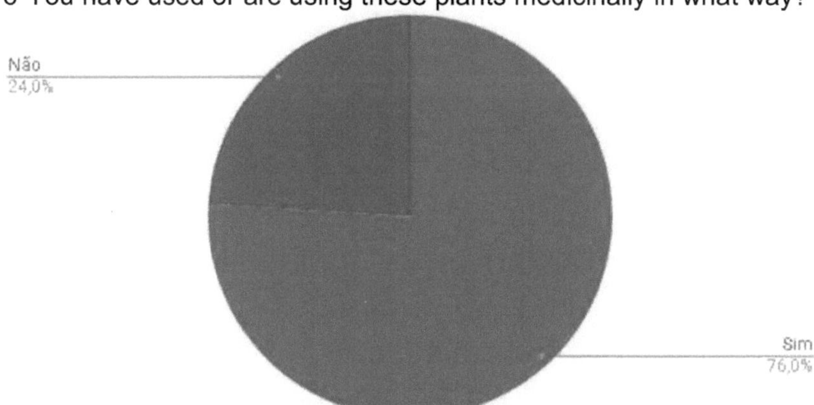

7-Do you believe that some plants can be bad for your health?

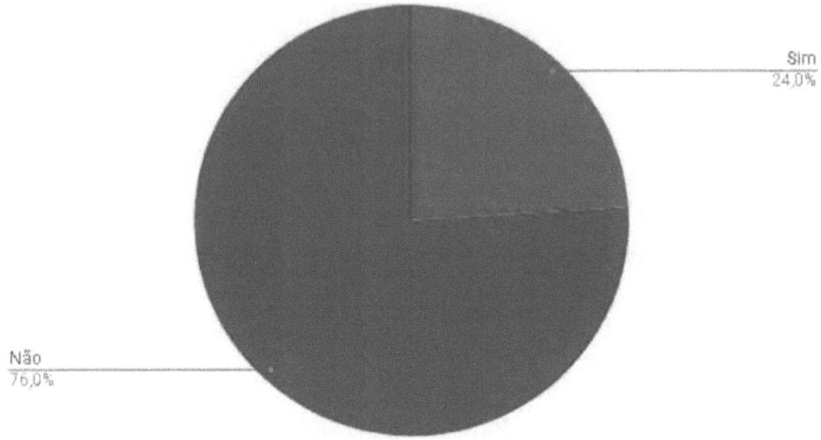

8-Do you know or can you identify plants that are toxic?

9-Do you have any plants that are restricted for use?

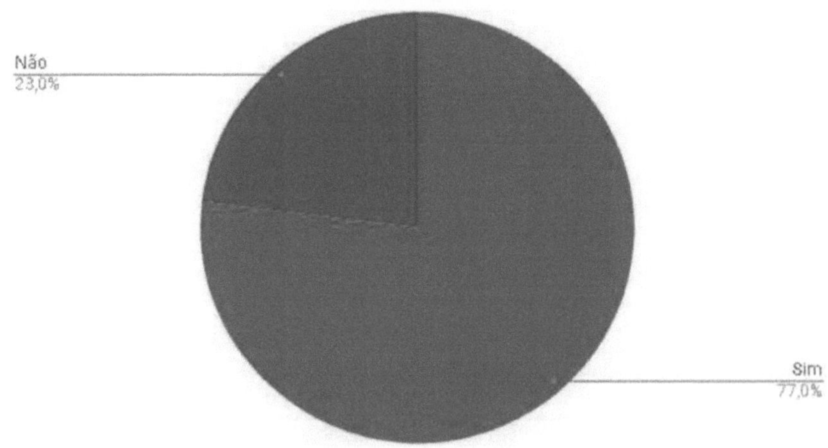

10-Do you believe that medicinal plants heal?

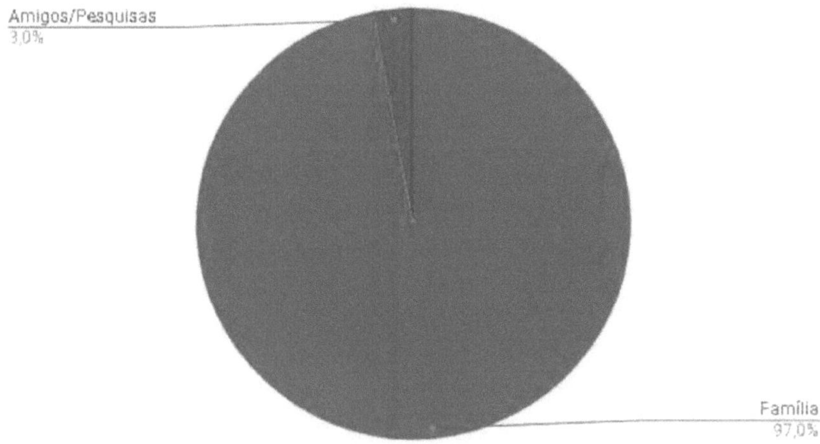

11-Who taught you how to use medicinal plants?

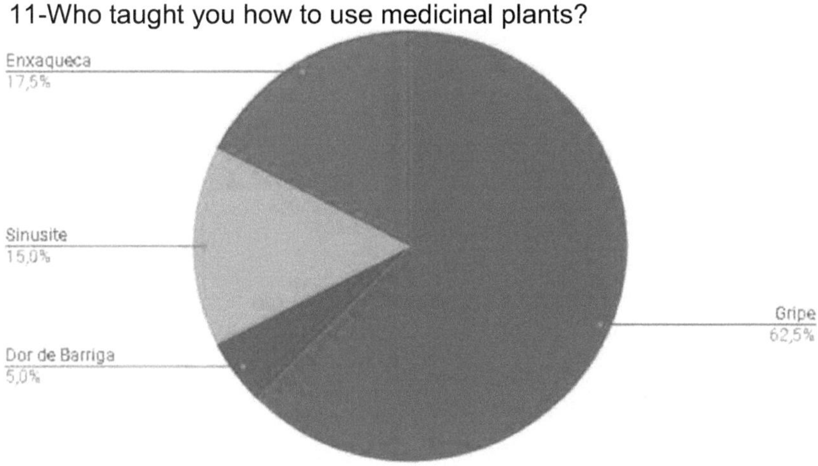

12-What are the most common health problems in your home?

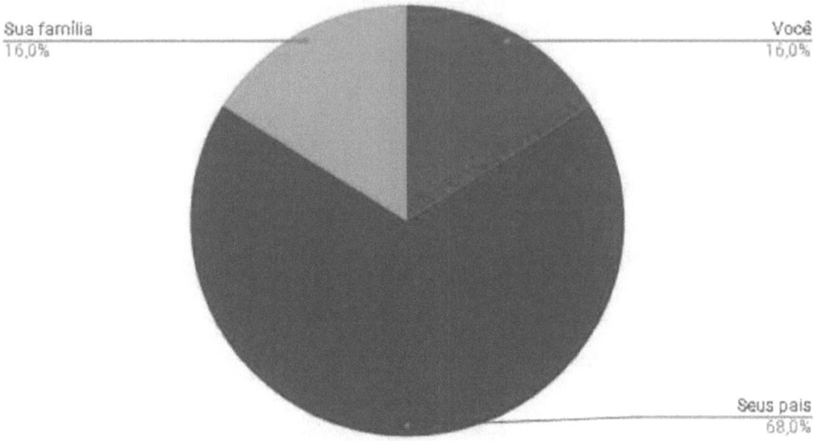

13-Who is in charge of solving these problems?

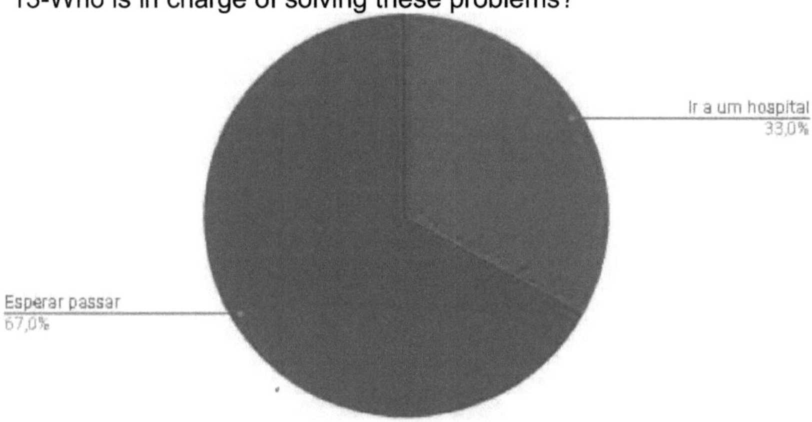

14-What is the first thing you do to treat a health problem?

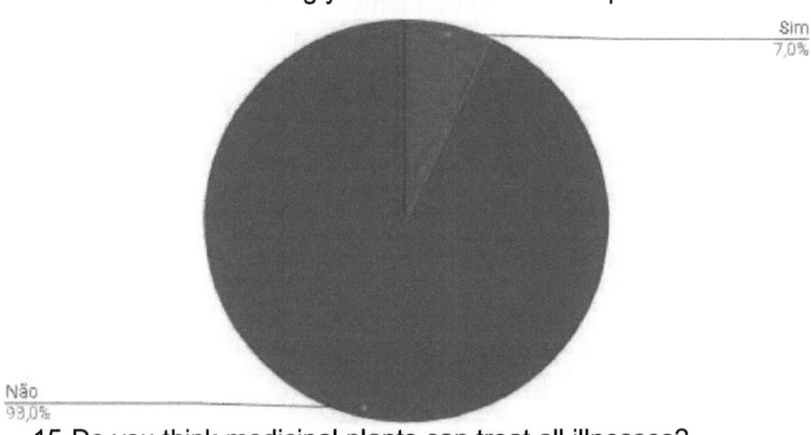

15-Do you think medicinal plants can treat all illnesses?

Shortly after developing the questionnaire with the students, a survey was also carried out with the civil servants, in order to finally make an average of which level of education and age knew more about medicinal plants and to analyse who used them the most, and it was also in which it was possible to notice different data to that of the students, which often generated concerns.

The staff survey was carried out at the IFSULDEMINAS Campus Muzambinho - MG, between June/2017 and November/2017. The civil servants who took part in the survey were chosen at random and on a voluntary basis, as was the case with the student survey. The interviews took place in the same way, using a semi-structured questionnaire with the same open and closed questions as the student questionnaire on the premises of the IFSULDEMINAS Campus Muzambinho - MG. The questionnaire consisted of questions about the medicinal species most frequently used by the interviewee and their family, who taught them how to use them and how they were prepared. The results obtained were subjected to descriptive statistical analysis using relative frequency. The interviews lasted around half an hour.

This survey interviewed 100 civil servants, showing that 79% of them said that they use medicinal plants casually, demonstrating that the majority of those interviewed use medicinal herbs.

Of the 100 people interviewed, 71% reported that they acquired their knowledge from their parents and grandparents, which is in line with ALMEIDA et al. (2009), who teaches that the use of medicinal plants is a tradition passed down from parents to children; few learnt to use traditional folk medicine with the help of the internet or friends.

Around 74 per cent of informants use them.

The interviewees who took part in the survey ranged in age from thirty-two to sixty. They mentioned three different species of medicinal plants that they used the most, the most frequently mentioned being Mint (Menta x villosa Huds) (33%), Lemon Balm (Cymbopogon citratus (DC). Stapf (25%) and Chamomile (Matricaria chamomilla) (15%).

It was found that 50 per cent of the interviewees use the species because of an illness, while the other 50 per cent said that they use medicinal herbs only because they have already acquired the habit or because they like it.

Another finding in the survey was that 76 per cent of civil servants believed that medicinal plants cured various illnesses, while only 24 per cent said that medicinal plants only cured illnesses when combined with other synthetic medicines.

The most commonly used methods for preparing the species were tea by decoction 50 per cent and tea by infusion 42 per cent, but the interviewees used other methods, and many of the interviewees boil the leaves to make the tea and don't just add them to the boiled water. Many also mentioned using other methods such as compresses and bottles, a type of remedy often used by older people.

It is worth pointing out that in relation to the toxicity of medicinal species, a worrying observation was noted, as well as such a worrying one in the students' survey, since 81 per cent of those interviewed said that they did not know how to identify which plants are toxic, and so they only use medicinal herbs recommended by friends and relatives. This reality is very serious, but very common among individuals from other communities, as the risk of poisoning can increase and occur frequently due to a lack of knowledge about the incorrect use of medicinal species and not recognising toxic species.

In terms of health problems, the most common and most frequent were

employees and their families was flu with 33%, sinusitis 30%, migraine with 20%, other

with 10% and stomach pain 7%. When some of these pathologies occur, 76 per cent of those interviewed say they wait until they have passed before seeking medical help, while 24 per cent say they seek help as soon as any of these symptoms occur.

The interviewees reported that those who use medicinal plants the most are the elderly 32.4%, adults 29.4%, children 23.5% and young people 14.7%. Below are the graphs of the results of the tabulated analyses of the answers given by the teachers interviewed.

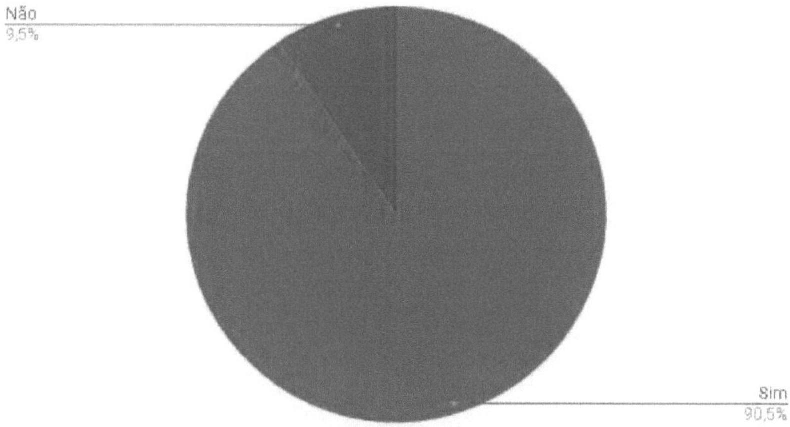

1) You have used or are using medicinal plants:

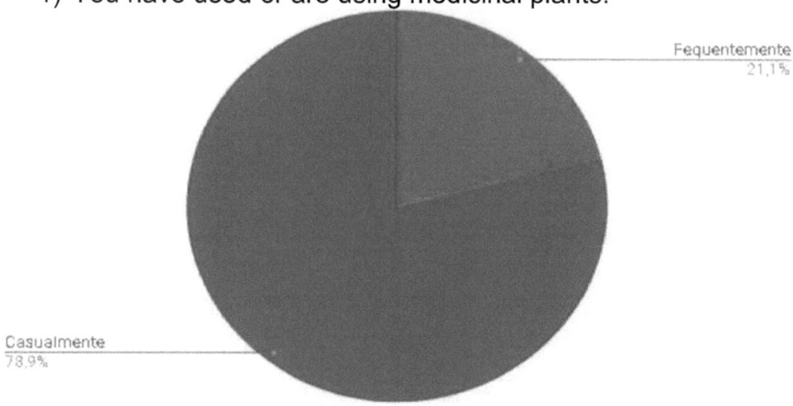

2) How often:

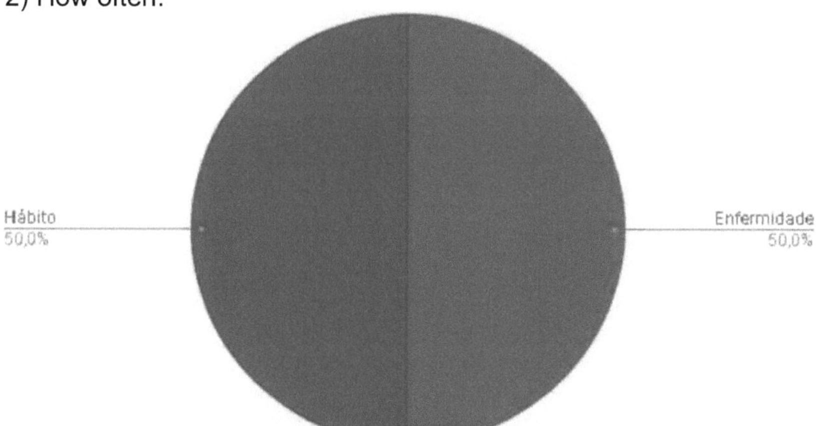

3) For what purpose do you use medicinal plants?

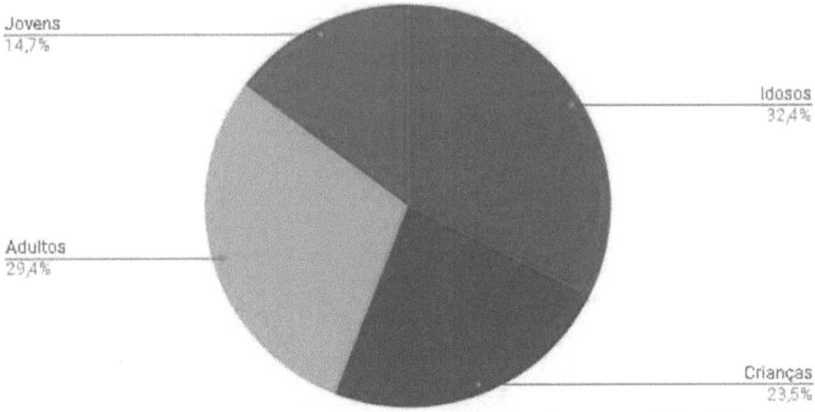

4) For you, which age group predominates in the use of Medicinal Plants:

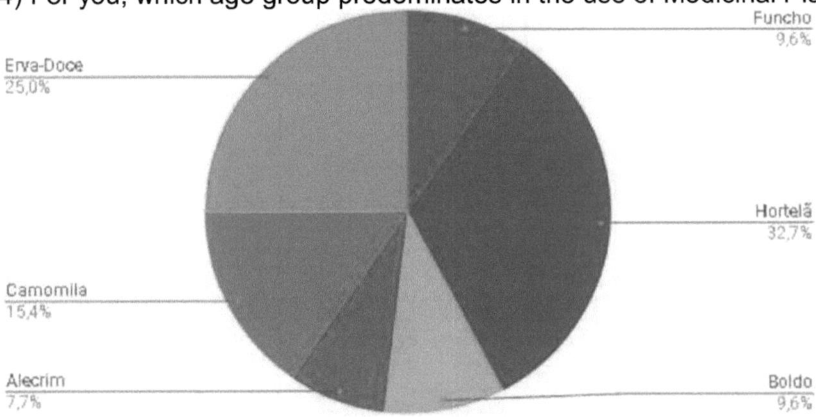

5)Which species do you usually use: x 5)Which species do you usually use:

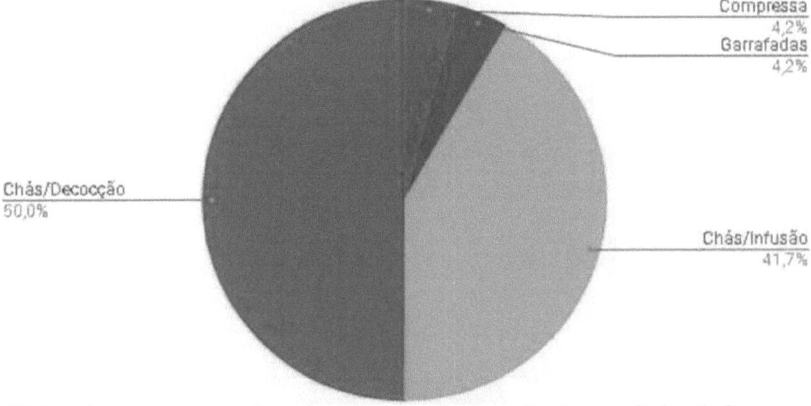

6)How have you used or do you use these plants medicinally?

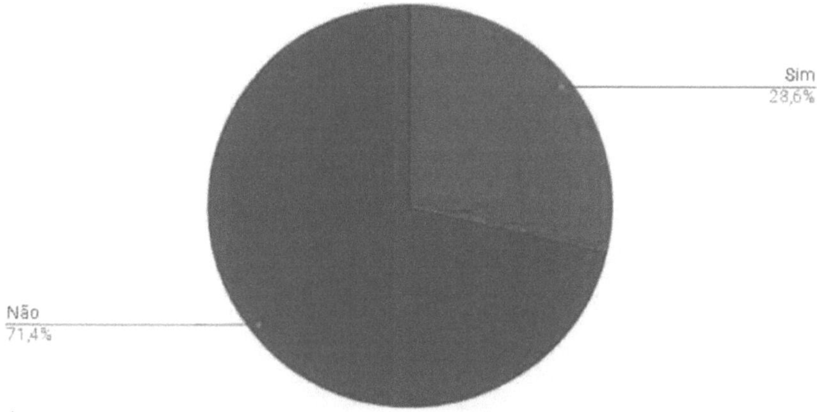

7)Do you believe that some plants can be bad for your health?

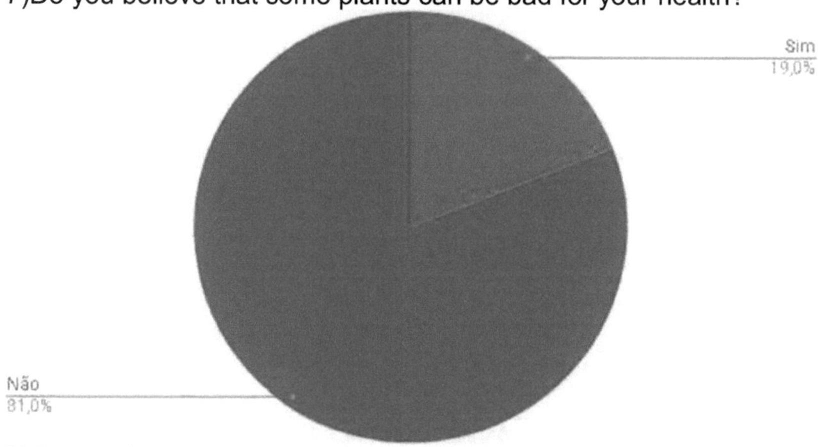

8) Do you know or can you identify plants that are toxic?

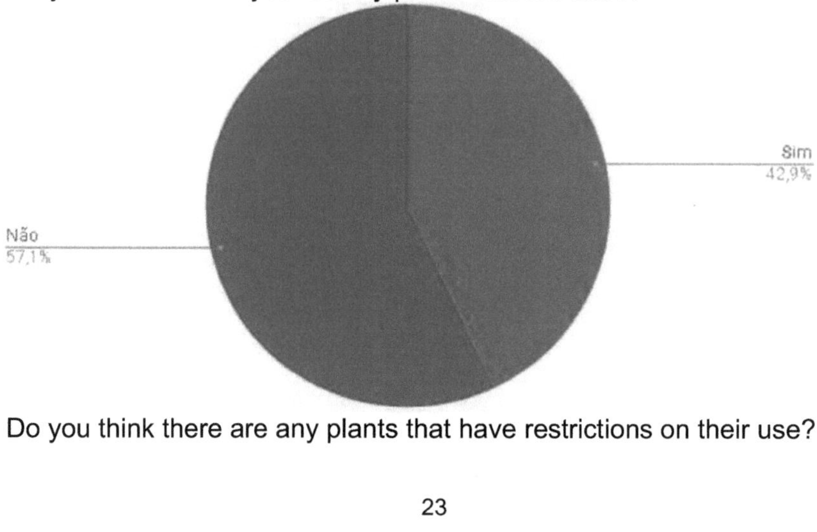

9) Do you think there are any plants that have restrictions on their use?

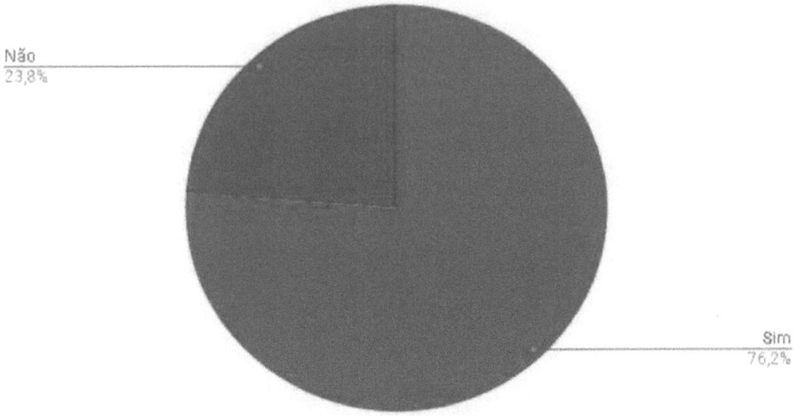

10) Do you believe that medicinal plants heal?

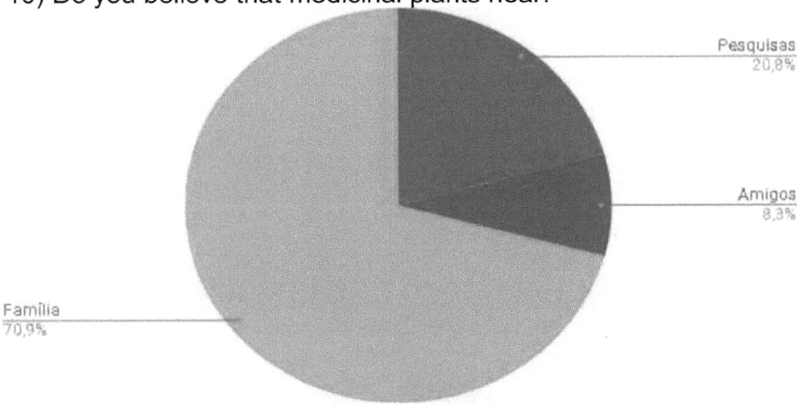

11) Who taught you how to use medicinal plants?

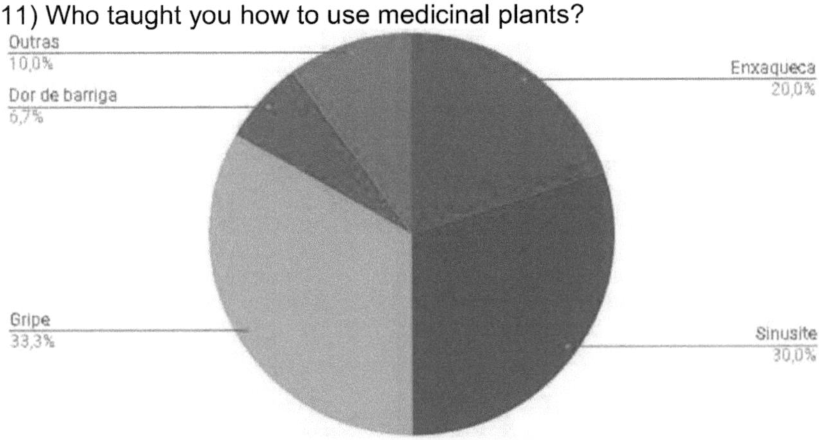

12) What are the most common health problems in your home?

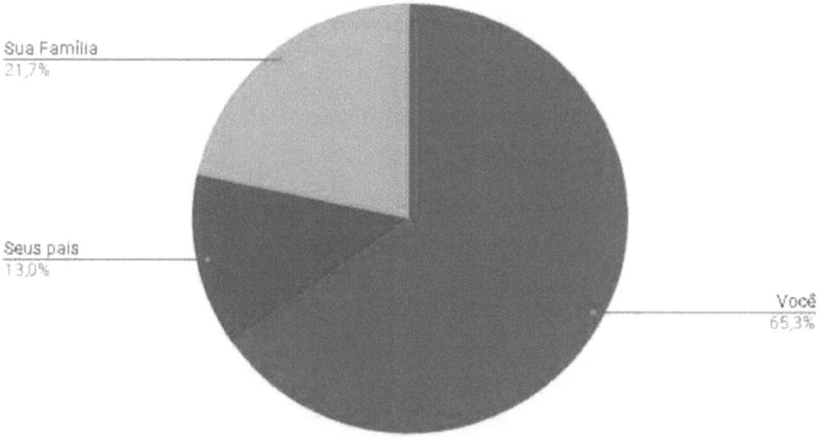

13) Who is in charge of solving these problems?

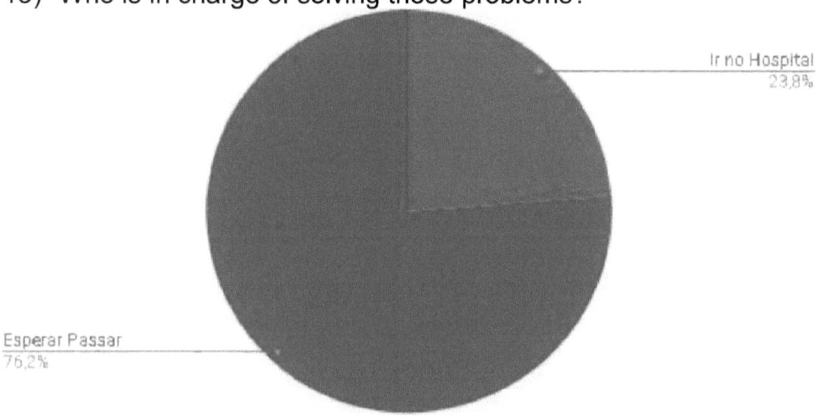

14) What is the first thing you do to treat a health problem?

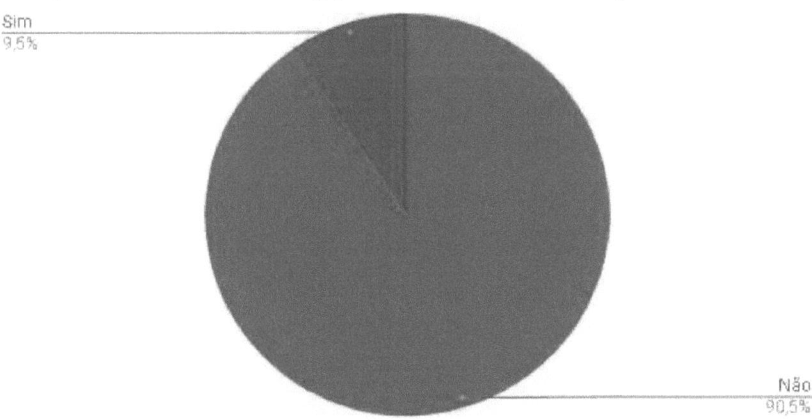

15) Do you think medicinal plants can treat all illnesses?

In order to demonstrate and compare the answers given by each person, graphs of all the answers given by the interviewees, both students and civil servants, were generated during the analyses of the tabulated information.Below are all the graphs of the analyses carried out.

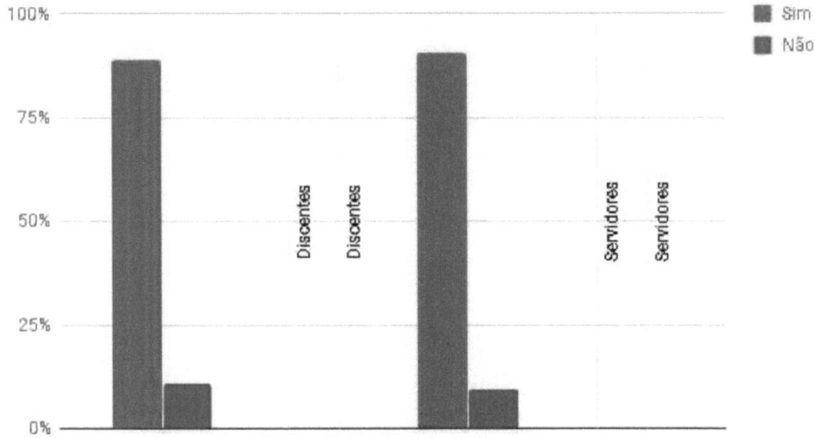

1)Have you ever used medicinal plants?

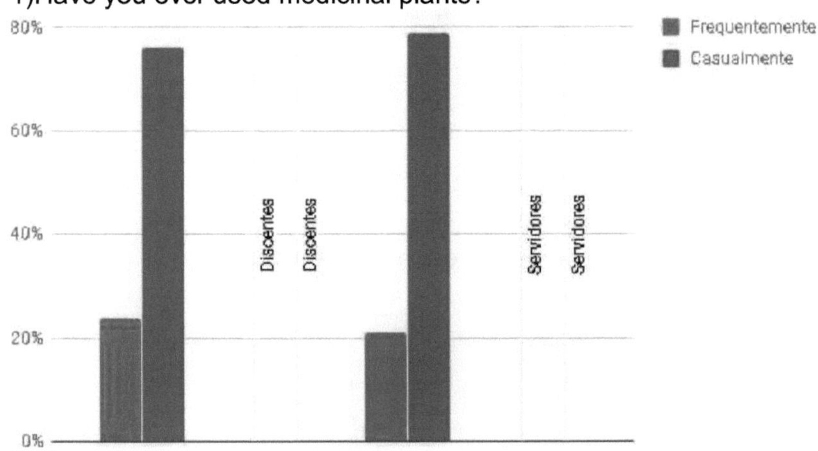

2)How often?

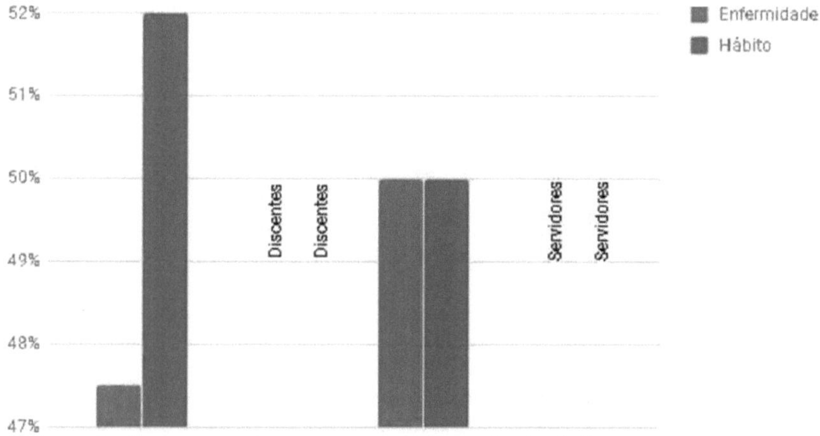

3)For what purpose do you use medicinal plants?

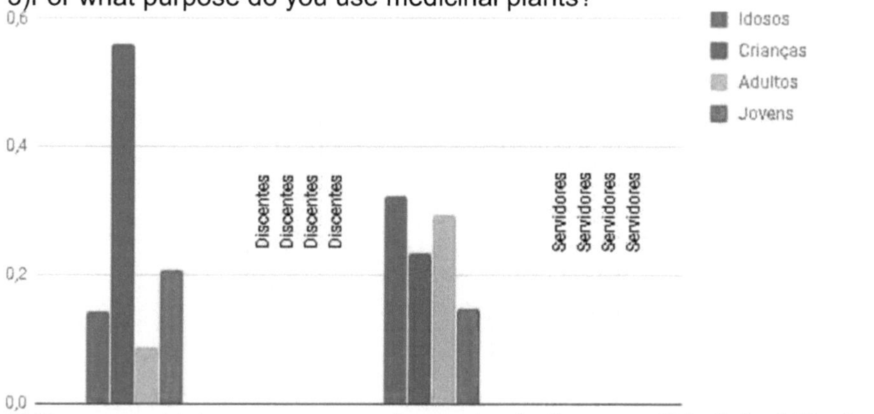

4) For you, which age group predominates in the use of Medicinal Plants:

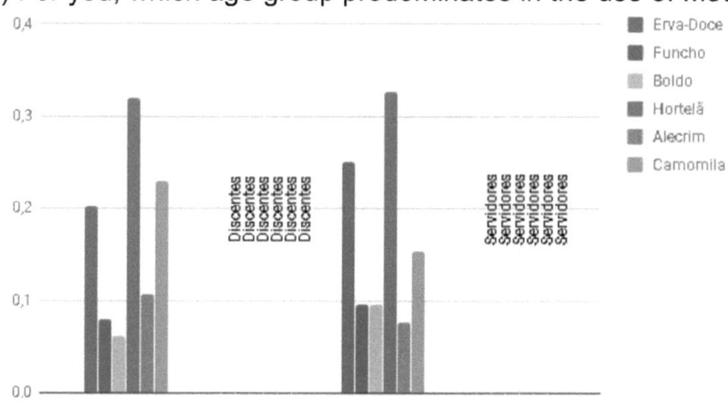

5)Which species do you use?

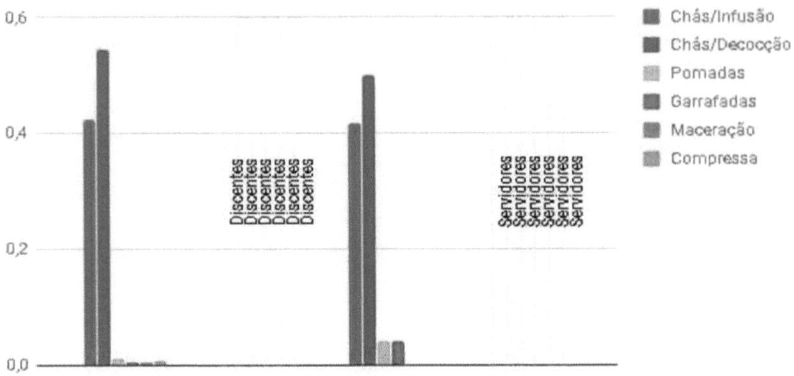

6)How have you used or do you use these plants medicinally?
saúde?

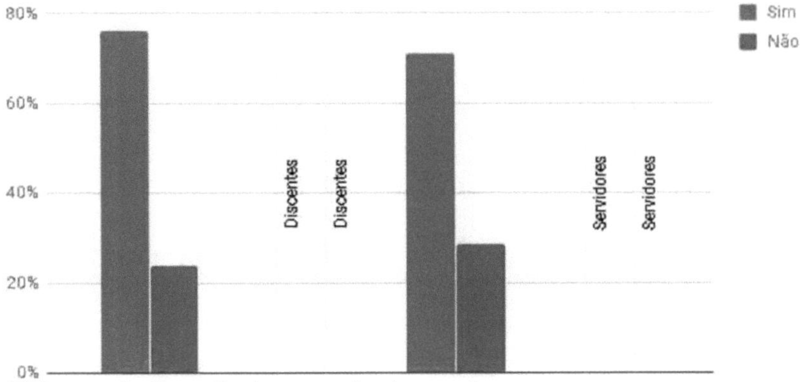

7)Do you believe that some plants can harm

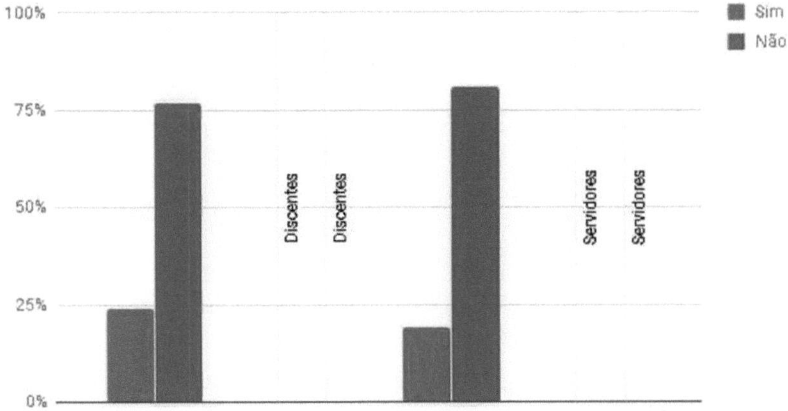

8) Do you know or can you identify plants that are toxic?

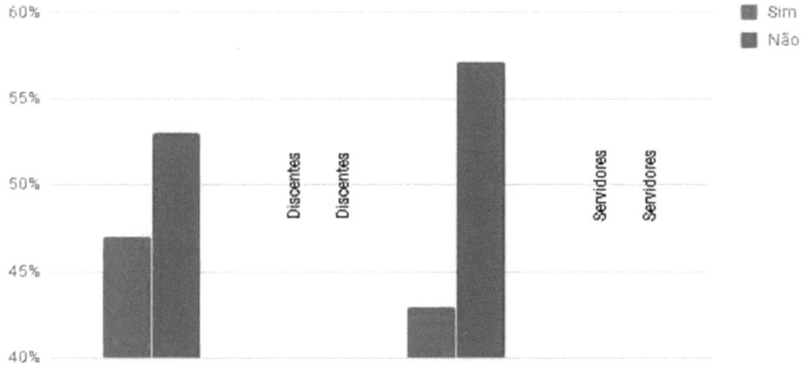

9) Are there any plants that you think are restricted from use?

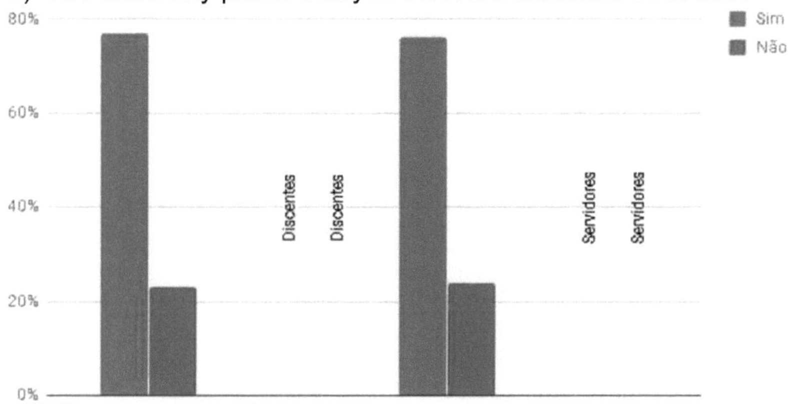

10) Do you believe that medicinal plants heal?

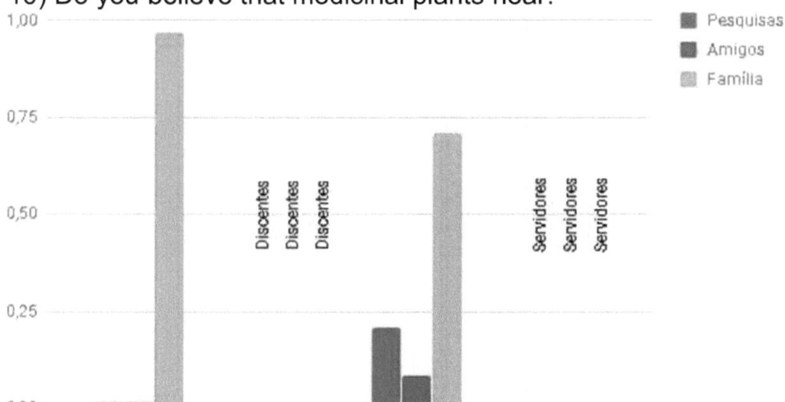

11) Who taught you how to use medicinal plants?

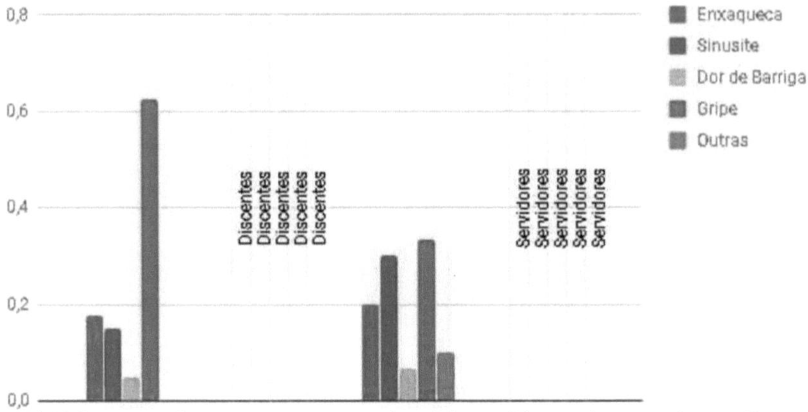

12) What are the most common health problems in your home?

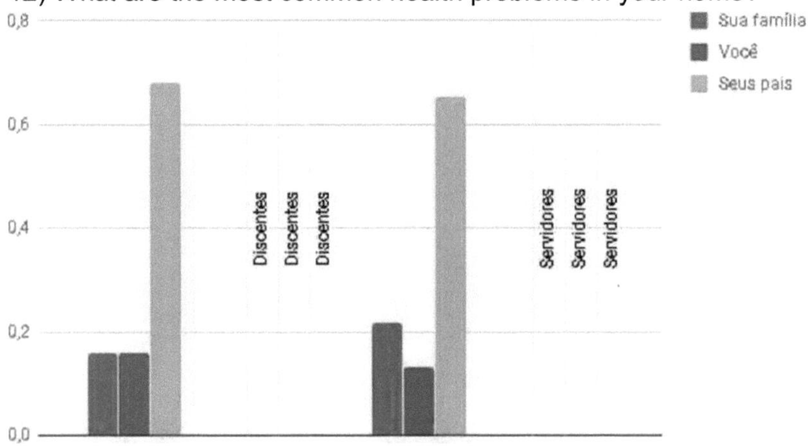

14) Who is in charge of solving these problems?

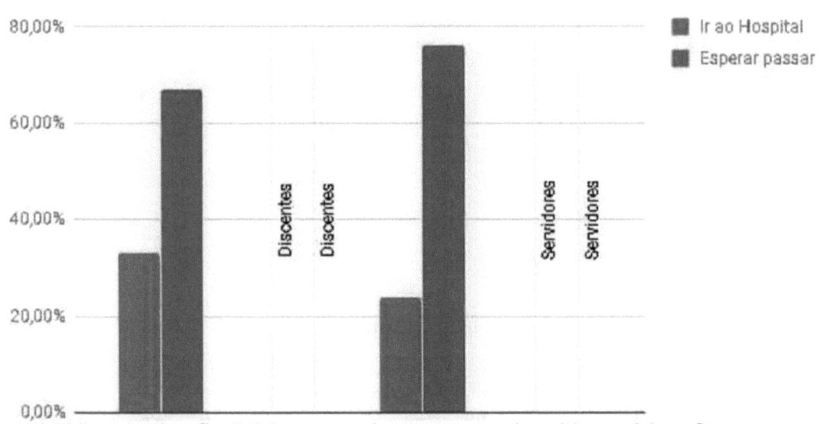

14) What is the first thing you do to treat a health problem?

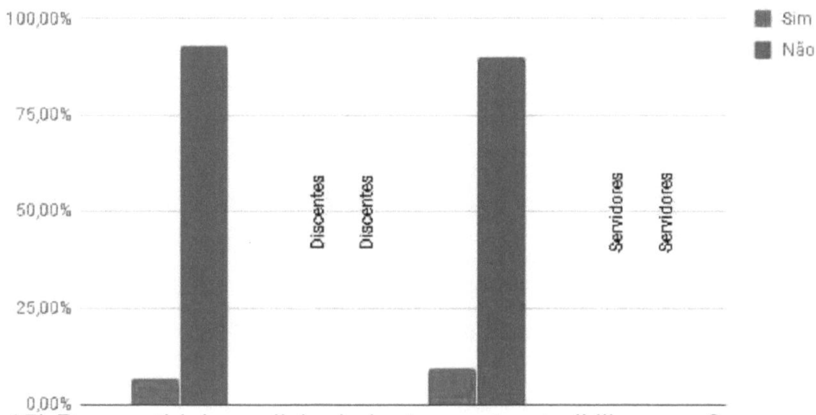

15) Do you think medicinal plants can treat all illnesses?

In these graphs, we can see that a lot of the information is similar and the data is close to each other in terms of percentage, and we can also see that even with the difference in age and education, the knowledge of a civil servant and a student is close.

After tabulating and generating all the graphs, the results were gratifying, and it was satisfying to know that the knowledge of traditional folk medicine, even with the passage of time, is being recovered and maintained, some of which in some regions of Brazil has already been lost.

CHAPTER 5

CONCLUSIONS

The species most used by students were *Mint (Menta x villosa* Huds) (32%), Chamomile *(Matricaria chamomilla)* (23%) and Lemon Balm *(Cymbopogon citratus* (DC) (20%). Stapf. and the most used by the civil servants were the same but in different percentages: Mint (Menta x villosa Huds) (33%), Lemon Balm (Cymbopogon citratus (DC) (25%). Stapf and Camomile
(Matricaria chamomilla) (15%).

The majority of students (77%) and civil servants (81%) interviewed do not know how to differentiate between medicinal and toxic species, using species recommended by friends and relatives and do not make rational use of the dosage of the species.

The most commonly cited methods of preparation were infusion and decoction, but there is a notable lack of knowledge regarding the recommended method of preparation for each medicinal species, and more study in this area and dissemination to both students and teachers is needed.

97% of the students interviewed and 71% of the civil servants interviewed use medicinal plants through knowledge passed down from their parents and grandparents.

Although they are young, the age range studied was between fifteen and twenty-five years old, they use medicinal plants frequently, receive this knowledge from their relatives and believe in their healing, although they know little about the toxicity of the species and therefore do not always make conscious use of medicinal herbs. As for the civil servants, the age range of those interviewed was between thirty-two and sixty years old. Most of them use medicinal plants casually and have received this knowledge from their relatives.

76 per cent of civil servants and 77 per cent of students believe in healing through medicinal plants, despite the fact that they know very little about the toxicity

of the species and therefore don't always make conscious use of medicinal herbs, running a serious risk of poisoning from them.

The most common health problems among civil servants and their families were flu (33%), sinusitis (30%), migraine (20%), stomach ache (7%) and others (10%). When some of these illnesses occur, 76% of those interviewed say they wait until they have passed before seeking medical help, while 24% say they seek help as soon as any of these symptoms occur. The most common health problems among students and their families were flu (62.5%), sinusitis (15%), migraine (17.5%) and stomach ache (5%). When these illnesses occur, 66% wait for them to pass and only 33% of the students interviewed go to a hospital to see what is really going on and how to treat it.

ACKNOWLEDGEMENTS

I would like to thank the Federal Institute of Education, Science and Technology of Southern Minas Gerais - Campus Muzambinho, NIPE (NÚCLEO INSTITUCIONAL DE PESQUISA E EXTENSÃO- NIPE) and COOPAM (COOPERATIVA ESCOLA DOS ALUNOS DA EAF DE MUZAMBINHO LTDA) for supporting the project. To the students interviewed who contributed to the data collection. To my family and friends for always supporting me.

ANNEX I

MEANS OF INTERVENTION TO OBTAIN DATA

Structured Questionnaire on Medicinal Plants:

Maria Vitória Pereira Vicente - Student of the Integrated Computer Technician Course at the Federal Institute of Education, Science and Technology Sul de Minas Gerais - Campus Muzambinho-MG -maviiipereira15@gmail.com

CONTEXTUALISATION

The purpose of the structured questionnaire is to verify the number of students who use medicinal plant species at the IFSULDEMINAS Campus Muzambinho - MG. A structured questionnaire was used to carry out this research.

SKILLS TO BE DEVELOPED DURING THE QUESTIONNAIRE PERIOD

- Awakening knowledge, interest, curiosity and information about medicinal plants through a questionnaire;
- To analyse the number of students who use medicinal plants and have knowledge of the area surveyed;
- Encouraging the use of medicinal herbs and the participation and co-operation of those interviewed in passing on the knowledge of traditional folk medicine;
- To make students aware of the importance of knowing the benefits and harms of medicinal plant species, and to teach them how to identify which plants are toxic.

KNOWLEDGE MOBILISED DURING THE QUESTIONNAIRE PERIOD

- Raising awareness of the importance of knowing medicinal plants in order to use them;
- Transmission of existing knowledge and new knowledge;
- Identification of the different methods of use of Medicinal Plants by students at the IFSULDEMINAS Campus Muzambinho - MG.

METHODOLOGICAL PROCEDURES DURING THE QUESTIONNAIRE PERIOD

A structured questionnaire on a sheet of paper (A4) containing fifteen questions related to the project on medicinal plants was used.

STRUCTURED QUESTIONNAIRE USED

Informant data:

Name:Address: ______________________________ Telephone:

____ Course:

____ Age or age group: ()15 to 20 ( ) 21 to 40 ( ) 41 to 60 ( ) 61 to 80 ( ) over 80.

1) You have used or are using medicinal plants:

( )Yes ( )No

2) How often:

( )Casually ( )Often

3) For what purpose do you use medicinal plants?

( )Due to having acquired the habit ( )Due to some illness

Another purpose:_____________________________________

4) For you, which age group predominates in the use of Medicinal Plants:

( )Children ()Youth ( )Adults ( )Seniors

5) Which species do you use?

( ) Fennel ( ) Boldo ( )Rosemary ( )Chamomile ( )Mint ( )Fennel

Other: ____

6) How have you used or do you use these plants medicinally?

( )Teas( )Decoction( )Infusion( )Maceration( )Bottled

()Compresses ( )Ointments

7)Do you believe that some plants can be bad for your health?

( ) Yes ( ) No

8)Do you know or can you identify plants that are toxic?

( ) Yes ( ) No

9)Do you think there are any plants that have restrictions on their use?

( ) Yes ( ) No

10)Do you believe that medicinal plants heal?

( ) Yes ( ) No

ll)Who taught you how to use medicinal plants?

12 What are the most common health problems in your home?

( ) Flu ( ) Bellyache ( ) Sinusitis ( ) Migraine

Other: ____

13)Who is in charge of solving these problems?

( ) You ( ) Your parents ( ) Your family

14)What is the first thing you do to treat a health problem?

( ) Go to a hospital/health centre ( )Wait for it to pass

15) Do you think medicinal plants can treat all illnesses?
( ) Yes ( ) No

EVALUATION OF THE QUESTIONNAIRE RESULTS

The evaluation will be carried out through the answers to the questionnaire, based on the interaction of the students during the interview, together with a survey of the information reported by the interviewees and obtained during the application of the questionnaire. The analyses were tabulated in a spreadsheet and then graphs were generated, which are included in the materials and methods section of this article.

BIBLIOGRAPHY

AGRA, M. F., FREITAS, P. F., BARBOSA-FILHO, J. M., Synopsis of the plants known as medicinal and poisonous in northeast of Brazil. Rev. Bras. Farmacogn., 17(1): 114-140, 2007.

ALBUQUERQUE, U.P.; LUCENA, R.F.P. Métodos e técnicas de pesquisa etnobotânica.Recife: NUPEEA, 2004. 189p.

ALMEIDA, N. F., et al. Ethnobotanical survey of medicinal plants in the city of Viçosa-MG. Rev. Bras. Farm., 90(4): 316-320, 2009.

AMOROZO, M. C.M. A abordagem etnobotânica na pesquisa de Plantas Medicinais.

In: Dl STATSI, L.C. (Org.). Medicinal Plants: Art and Science, a study guide

interdisciplinary. São Paulo: EDUSP. p. 47-68. 1996.

BUCHILLET, D. (Org.) Traditional medicines and Western medicine in the Amazon. CABALLERO, J. Perspectiva para el quehacer etnobotanico en México. Belém: Ed. Cejup.l991.p.63-64.

CAMARGO, M.T.L.. Popular medicine. Rio de Janeiro: Campaign for the Defence of Brazilian Folklore, 46p., 1976.

DIEGUES, A. C. S. The modern myth of untouched nature. HUCITEC, São Paulo. 1996.

DORIGONI, P. A. et al. Data survey on medicinal plants in popular use in the municipality of São João do Polêsine, RS, Brazil. I - Relationship between diseases and species used. Revista Brasileira de Plantas Medicinais, v. 4, n. l,p. 69-79, 2001.

ELISABETSKY, E. Ethnopharmacology of some Brazilian tribes. In: RIBEIRO, D. Suma etnológica brasileira. Petrópolis: Vozes, 1997.

FRANCO, A. L. P., et al. Evaluation of the chemical composition and antibacterial activity of the essential oils of *Aloysia gratíssima* (Gillies & Hook) Trone. (ALFAZEMA), *Ocimum gratissimum L.,* (ALFAVACA-CRAVO) and *Curcuma longa* L., (AÇAFRÃO). Rev. Eletr. de Farm., 4(2): p. 208-220, 2007.

GUARIM NETO, G., SANTANA, S. R. & SILVA, J. V. B.. 2000. Etonobotanical notes on species of Sapiendaceae jussieu. Acta Bot. Bras. v.14 n.3 São Paulo Sept./Dec. 2000.

HAMILTON, A. Medicinal plants and conservation: issues and approaches. [s.l.]: International Plants Conservation Unit, 2003.

MARTINS, E. R., et al. Medicinal Plants. Viçosa: Ed. UFV, 220 p., 2003.

OLIVEIRA, C.J.; ARAÚJO, T.L. Plantas medicinais: usos e crenças de idosos portadores de hipertensão arterial. Revista Eletrónica de Enfermagem, v. 9, n. 1, p. 93-105, 2007.

PEREIRA, N. F., NAKAMURA-PEREIRA, S. M., Contribution to the study of toxic plants and their antagonists: rat grass, the Rubiaceae, Palicourea macgravi ST. Hill. Rev. Bras. Farm., Rio de Janeiro, (87): p. 60-62, 2005.

PERES, Cristina Silveira; MARQUES, Hetiene Pereira. Ethnobotanical survey of medicinal plants in the rural community of Mato Dentro, Campos Gerais - MG. 2010. 38 f. TCC (Graduation) - Biological Sciences Course, Campos Gerais Faculty of Sciences and Health, Campos Gerais, 2010.

PILLA, M. A. C.; AMOROZO, M. C. M.; FURLAN, A. Obtaining and using medicinal plants in the Martim Francisco district, Municipality of Mogi-Mirim, SP, Brazil. Acta Botânica Brasílica, v. 20, n. 4, p. 789-802, 2006.

RODRIGUES, A.G.; CASALI, V.W.D. - Medicinal plants, popular knowledge and ethnoscience. In: RODRIGUES, A.G.; ANDRADE, F.M.C.; COELHO, F.M.G et al. - Medicinal and aromatic plants: ethnoecology and ethnopharmacology. Viçosa: UFV, p. 25-76, 2002

SILVA, S. R. S., et al. Analysis of the chemical constituents and antimicrobial activity of the essential oil of Melaleuca altemifolia Cheel. Rev. Bras. Pl. Med., Botucatu, 6(1): p. 63-70, 2003.

TUROLLA, M.S.; NASCIMENTO, E.S. Toxicological information on some herbal medicines used in Brazil. Revista Brasileira de Ciências Farmacêuticas, v. 42, p. 289-306, 2006.

MIX
Papier aus verantwortungsvollen Quellen
Paper from responsible sources
FSC® C105338

Printed by Books on Demand GmbH, Norderstedt / Germany